Abdelhafid Mimouni

The Silent Weight of Hypertension

AF300650

Abdelhafid Mimouni

The Silent Weight of Hypertension

ScienciaScripts

Imprint

Any brand names and product names mentioned in this book are subject to trademark, brand or patent protection and are trademarks or registered trademarks of their respective holders. The use of brand names, product names, common names, trade names, product descriptions etc. even without a particular marking in this work is in no way to be construed to mean that such names may be regarded as unrestricted in respect of trademark and brand protection legislation and could thus be used by anyone.

Cover image: www.ingimage.com

This book is a translation from the original published under ISBN 978-620-6-71566-5.

Publisher:
Sciencia Scripts
is a trademark of
Dodo Books Indian Ocean Ltd. and OmniScriptum S.R.L publishing group

120 High Road, East Finchley, London, N2 9ED, United Kingdom
Str. Armeneasca 28/1, office 1, Chisinau MD-2012, Republic of Moldova, Europe
Printed at: see last page
ISBN: 978-620-7-90348-1

Copyright © Abdelhafid Mimouni
Copyright © 2024 Dodo Books Indian Ocean Ltd. and OmniScriptum S.R.L publishing group

The Silent Weight of Hypertension

Author: Dr. Abdelhafid Mimouni is an independent researcher specializing in bioinorganic systems chemistry, with extensive expertise in macromolecular synthesis and characterization. He obtained his PhD in chemistry from the University of Paris XII in 1997 and a Diplôme d'Études Approfondies in bioinorganic systems from the University of Paris XI in 1993.

Summary:

This work examines the crucial role of adrenal hormones, such as cortisol and aldosterone, in a variety of essential physiological processes. Focusing on their influence on the health of the elderly, we explore how these hormones interact with other regulatory systems to maintain the body's equilibrium. We also examine age-associated changes in adrenal gland function and their impact on overall health. In addition, we discuss therapeutic implications and advances in bioinorganic research to treat disorders related to impaired adrenal function. This work aims to deepen our understanding of the underlying mechanisms and propose future perspectives for personalized medicine in aging and health.

Plan

Introduction

Adrenal hormones, secreted by the adrenal glands, play an essential role in regulating various physiological processes crucial to the maintenance of human health. Central to these processes are elements such as stress response, metabolism, immune function and blood pressure regulation. These hormones, including cortisol and aldosterone, interact in complex ways with other regulatory systems in the body to coordinate these vital functions.

This research explores in depth the role of adrenal hormones in the context of aging, highlighting their influence on the health of older individuals. In this introduction, we outline the major themes that will be developed in this work. We'll start with a discussion of the main functions of adrenal hormones, then move on to the changes associated with aging in adrenal gland function. We will then highlight the importance of understanding these mechanisms in order to develop therapeutic approaches adapted to the elderly.

Finally, we will briefly outline the structure of this research, describing the different chapters and specific aspects that will be addressed. The aim of this work is to deepen our understanding of the role of adrenal hormones in aging, and to open up new perspectives for the health management of older individuals.

Chapter 1: Introduction to the physiology of the adrenal glands

The discovery and development of corticoids represents an important chapter in the history of modern medicine. These compounds were initially isolated and identified during the 20th century, ushering in a new era in the treatment of a variety of medical conditions. Here's a detailed exploration of the key milestones in this history:

1	Early discoveries : The first discoveries linked to corticoids date back to the early 20th century, when researchers began to identify and isolate compounds from the adrenal glands. In 1936, Edward Kendall and his colleagues succeeded in extracting a hormonal compound from the adrenal glands, which they named "cortisone".

2	Synthesis and mass production: Although cortisone was the first corticoid hormone to be isolated, its production in sufficient quantities for medical use was initially limited. However, in the years that followed, major advances in chemical synthesis enabled the mass production of corticoids, paving the way for their widespread use in medical practice.

3	Clinical applications: In the 1940s and 1950s, corticoids began to be used in the treatment of a variety of medical conditions, from inflammatory and autoimmune diseases to allergies and endocrine disorders. Their anti-inflammatory and immunosuppressive properties made them invaluable in the treatment of many serious conditions.

4	Subsequent research: Over the decades, research into corticoids has continued, deepening our understanding of their mechanism of action, their effects on the body and their clinical implications. Researchers from all over the world have contributed to this growing understanding, opening up new perspectives on the use and therapeutic potential of corticoids.

5   Recent developments and future prospects: In the 21st century, corticosteroids continue to be widely used in medical practice, but extensive research is still underway to explore new applications, refine treatment protocols and minimize the side effects associated with their use. Advances in corticosteroid research offer promising insights into their potential for treating a variety of conditions, while underlining the continuing importance of innovation and clinical   vigilance in their use.

Corticosteroids are an interesting subject. They are often used in the medical field for their anti-inflammatory and immunosuppressive effects. These drugs mimic the corticosteroid hormones produced naturally by the body's adrenal glands.

There are two main types of corticosteroid: glucocorticoids and mineralocorticoids. Glucocorticoids are often prescribed for their anti-inflammatory effect, while mineralocorticoids are mainly involved in regulating electrolyte balance and blood pressure.

Corticosteroids are used to treat a variety of medical conditions, including allergies, asthma, arthritis, autoimmune diseases, dermatological disorders, respiratory disorders, and many others.

However, prolonged use or use in high doses can lead to a variety of side effects, including increased risk of infection, reduced bone density, weight gain, increased blood pressure, and metabolic disorders such as diabetes.

In addition, abruptly stopping corticosteroids after prolonged use can lead to a withdrawal syndrome with symptoms such as fatigue, muscle and joint pain, and sometimes a relapse of symptoms of the condition being treated.

Because of these potential side effects, it's important that the use of corticosteroids be supervised by a healthcare professional, and that patients be closely monitored during treatment.

Mineralocorticoids are a class of steroid hormones produced by the adrenal gland, mainly by the outer layer called the glomerular zone. Aldosterone is the best-known example of a mineralocorticoid.

Their main physiological role is to regulate the balance of electrolytes, particularly sodium (Na+) and potassium (K+), in the body. Here's how they work to regulate electrolyte balance and blood pressure:

1   Sodium reabsorption: mineralocorticoids, mainly aldosterone, act on the kidneys to promote sodium reabsorption in the renal tubules. This means that more sodium is retained in the body rather than being excreted in the urine. Sodium is an essential electrolyte involved in many physiological processes, including maintaining water balance and regulating osmotic pressure.

2   Potassium excretion: as well as promoting sodium reabsorption, mineralocorticoids also increase potassium excretion in the kidneys. This helps maintain an appropriate potassium balance in the body. Excess or deficiency of potassium can have serious consequences for cardiac and muscular function.

3.   water retention: when sodium is reabsorbed in the kidneys, water tends to follow by osmosis. Thus, the action of mineralocorticoids on sodium reabsorption also contributes to water retention in the body, which can increase blood volume and thus blood pressure.

By regulating electrolyte balance and water retention, mineralocorticoids play a crucial role in maintaining blood pressure. Mineralocorticoid dysfunction can lead to electrolyte disorders and variations in blood pressure, which in turn can lead to health problems such as hypertension or hypokalemia.

Let's dive deeper into the relationship between mineralocorticoids, electrolytes and blood pressure, focusing on the consequences of dysfunction.

1.   arterial hypertension :

        oMineralocorticoids, particularly aldosterone, promote sodium retention in the kidneys, resulting in increased blood volume.

        oIncreased blood volume leads to increased pressure on blood vessel walls, resulting in higher blood pressure.

oMineralocorticoid dysfunction, such as excessive aldosterone production, can lead to excessive sodium retention and increased blood pressure, contributing to the development of hypertension.

2.   hypokalemia :

oMineralocorticoids also increase potassium excretion in the kidneys.

oMineralocorticoid dysfunction, such as excessive aldosterone production, can lead to excessive potassium loss.

oLow blood potassium levels, or hypokalemia, can have serious consequences, including abnormal heart rhythms, muscle weakness and neurological disorders.

oHypokalemia can also contribute to hypertension by altering the function of smooth muscle cells in blood vessel walls, which can lead to abnormal constriction of blood vessels and increased blood pressure.

In summary, mineralocorticoid dysfunction can disrupt the balance of electrolytes, particularly sodium and potassium, leading to increased blood pressure (hypertension) and abnormally low levels of potassium in the blood (hypokalemia). These conditions can lead to a range of serious health problems, and often require medical treatment to control and manage effectively.

Prednisolone is a glucocorticoid, a class of corticosteroids. Glucocorticoids are widely used in medical practice for their powerful anti-inflammatory and immunosuppressive effects. Prednisolone is derived from prednisone, another corticosteroid, and is often prescribed to treat a variety of inflammatory, allergic and autoimmune conditions, such as asthma, rheumatoid arthritis, inflammatory bowel disease, skin diseases, and many others.

Glucocorticoids, including prednisolone, can affect blood pressure (BP) balance, although their effect is generally less pronounced than that of mineralocorticoids.

Here's how prednisolone can affect blood pressure:

1.   inflammatory response: prednisolone, as a glucocorticoid, suppresses inflammation in the body. In some cases, chronic inflammation can contribute to high blood pressure. By suppressing this inflammation, prednisolone can indirectly help reduce blood pressure in some people.

2.   sodium and water retention: although glucocorticoids are not as potent as mineralocorticoids in sodium and water retention, high doses or prolonged use may result in mild sodium and water retention, which may contribute to an increase in blood pressure in some susceptible individuals.

3   Effect on carbohydrate metabolism: Glucocorticoids can increase insulin resistance and promote lipogenesis, which can lead to increased blood glucose levels and possibly contribute to the development of hypertension in some people, particularly those with a predisposition to diabetes or glucose intolerance.

However, it is important to note that the effect on blood pressure of prednisolone and other glucocorticoids may vary from person to person, depending on factors such as dose, duration of treatment, individual sensitivity, and the presence of other underlying medical conditions. It is therefore essential that the use of prednisolone be carefully monitored by a healthcare professional, especially in patients with a history of high blood pressure or other cardiovascular problems.

In a person with a blood pressure (BP) of 12 (which is normal blood pressure) and an age of 57, taking prednisolone can potentially affect blood pressure in several ways:

1.   slight increase in blood pressure: Prednisolone may have a minimal effect on increasing blood pressure, mainly due to its ability to retain sodium and water in the body. However, this increase is generally slight and may not be clinically significant in all individuals.

2   Individual variation: The effect of prednisolone on blood pressure may vary from person to person.   Some people may be more sensitive to these effects and see a more noticeable increase in their blood pressure, while others may not be significantly affected.

3   Necessary monitoring: Since you have normal blood pressure, it is important to monitor your blood pressure closely if you start taking prednisolone. Your doctor may recommend regular blood pressure checks to detect any significant increase and adjust treatment if necessary.

4.   other risk factors: It's also important to consider other risk factors for cardiovascular disease, such as diabetes, obesity, smoking and family history of heart disease, as these factors may influence how prednisolone affects your overall cardiovascular health.

In summary, although prednisolone may have a minimal effect on increasing blood pressure in some people, it is essential to discuss any potential changes with your doctor and to monitor your blood pressure closely while taking this medication.

Prednisone, like prednisolone, belongs to the glucocorticoid class of steroid hormones naturally produced by the adrenal glands. Its anti-inflammatory mechanism is complex, involving several actions at cellular and molecular level. Here's a simplified explanation of prednisone's anti-inflammatory mechanism:

1   Inhibition of inflammatory mediator synthesis: Prednisone acts at cellular level to inhibit the synthesis of several inflammatory mediators, such as prostaglandins, leukotrienes and pro-inflammatory cytokines. These mediators play a key role in triggering and propagating the inflammatory response.

2   Suppression of immune cell activity: Prednisone suppresses the activity of immune cells, notably T and B lymphocytes, macrophages and monocytes. These cells are involved in the immune response and the release of pro-inflammatory cytokines. By inhibiting their function, prednisone reduces the inflammatory response.

3   Inhibition of immune cell migration: Prednisone reduces the ability of immune cells to migrate to sites of inflammation. This reduces the accumulation of immune cells in inflamed tissues, helping to alleviate inflammation.

4   Stabilizing cell membranes: Prednisone can also stabilize cell membranes, reducing the release of certain inflammatory mediators.

5	Decreased vascular permeability: Prednisone can reduce the permeability of blood vessels, thereby reducing the leakage of fluid and inflammatory cells into surrounding tissues.

Overall, prednisone exerts its anti-inflammatory action by complexly modulating the immune response and reducing the production and activity of inflammatory mediators. However, it is important to note that prolonged or high-dose use of prednisone can lead to undesirable side effects, including immune system suppression, osteoporosis, high blood pressure and other health problems. Therefore, its use should be closely monitored by a healthcare professional.

One theory seems interesting and addresses several important aspects of human physiology related to aging and adrenal gland function:

1	Aging and adrenal function: It is well established that adrenal gland function declines with age. The adrenal glands produce hormones such as cortisol, which plays a crucial role in regulating the immune response and metabolism. Over time, this hormone production capacity can diminish, which can affect immune response and blood pressure regulation.

2	Age-related hypertension and hypotension: Aging is often associated with changes in blood pressure. Some older people develop hypertension, while others may experience hypotension. Hormonal imbalances, including those involving adrenal hormones such as cortisol, can contribute to these changes in blood pressure.

3   Role of corticosteroids in treatment : Corticosteroids, such as prednisone or prednisolone, are often used to treat a variety of conditions related to impaired adrenal function, as well as other inflammatory and autoimmune disorders. However, their use must be carefully monitored because of their potential side effects, notably on the immune system and blood pressure regulation.

4.   Future research: Further exploring the link between aging, adrenal gland function, immune response and blood pressure could be a promising area of research. Understanding how these processes interact may help develop better treatment strategies for the elderly, while minimizing the side effects of drugs such as corticosteroids.

In summary, your theory raises important questions about how aging affects adrenal gland function, immune response and blood pressure, as well as the role of corticosteroids in treating these changes. Exploring these ideas further could contribute to a better understanding of the underlying mechanisms and to new therapeutic approaches to improve the health of the elderly.

Chapter 2: Aging and adrenal function

The aging process brings with it a series of physiological changes in the body, including altered adrenal gland function. In this chapter, we will explore in detail the effects of aging on adrenal function and the implications of these changes for the health of the elderly.

Analysis of age-related changes in adrenal gland function

Adrenal gland function undergoes significant changes with aging, which can have a significant impact on the overall health of the elderly. This analysis of age-associated changes in adrenal gland function will provide a better understanding of the implications of these alterations on the health of aging individuals.

Structural and functional alterations: With age, the adrenal glands undergo structural and functional alterations. These changes include a reduction in the size of the adrenal glands and an alteration in cellular composition, notably a decrease in the number of hormone-producing cells. These structural alterations can compromise the adrenal glands' ability to efficiently produce and release hormones into the bloodstream.

Impaired hormone production: Decreased adrenal function in the elderly is often associated with impaired hormone production, particularly of cortisol. Cortisol is a key hormone produced by the adrenal glands, involved in stress response, metabolic regulation and modulation of the immune response. Decreased cortisol production can lead to imbalances in these physiological processes, contributing to the development of various medical conditions.

Impact on stress response: Cortisol is essential for regulating the stress response. Reduced cortisol production can lead to an altered stress response in the elderly, making them potentially more vulnerable to the negative health effects of stress, such as anxiety, depression and cardiovascular disease.

Effects on metabolism: Cortisol also plays a crucial role in regulating metabolism, including the management of sugars and fats in the body. Reduced cortisol production can disrupt this metabolic balance, increasing the risk of developing type 2 diabetes, abdominal obesity and other metabolic disorders in the elderly.

Implications for overall health: Alterations in adrenal function in the elderly can have a significant impact on their overall health, contributing to the development of common age-related diseases such as cardiovascular disease, diabetes, osteoporosis and cognitive impairment.

In summary, analysis of age-related changes in adrenal gland function highlights the importance of these alterations for the health of the elderly. A thorough understanding of these alterations will enable the development of disease prevention and management strategies for the elderly, aimed at promoting their long-term health and well-being.

Consequences of reduced adrenal function on the health of the elderly

Reduced adrenal function in the elderly can lead to orthostatic hypotension, a condition in which blood pressure falls sharply when the person stands up from a sitting or lying position. This drop in pressure can lead to dizziness, lightheadedness and even fainting, increasing the risk of falls and injuries in the elderly.

Studies have shown that genetics can play a role in predisposition to orthostatic hypotension. For example, certain genetic mutations can alter blood pressure regulation, increasing susceptibility to this disorder in the elderly. In addition, research has identified genetic variants associated with reduced levels of certain adrenal hormones, which may contribute to altered stress response and blood pressure regulation problems in the elderly.

While genetics may play a role in predisposing to this disorder, other factors, such as lifestyle and co-morbidities, may also contribute to its development. A holistic approach to the health of the elderly, taking into account both genetic and environmental factors, is essential to effectively prevent and manage these health problems.

Chapter 3: Immunomodulation by adrenal hormones

This chapter delves into the complex interaction between adrenal hormones and the immune system, highlighting the crucial role of cortisol and other hormones in regulating the immune response.

The central role of cortisol and adrenal hormones in balancing the immune response

Cortisol, often referred to as the stress hormone because of its primary role in regulating stress responses, reveals a much broader and more complex function in the field of immunology. This hormone, along with other adrenal hormones such as aldosterone and DHEA, act as key modulators of the immune system, orchestrating a cascade of physiological responses crucial to the body's protection.

Through sophisticated mechanisms, cortisol finely regulates both innate and adaptive immune responses. For example, it influences the production of cytokines, the signaling molecules that coordinate the immune response, by inhibiting the synthesis of pro-inflammatory cytokines and promoting the synthesis of anti-inflammatory cytokines. Cortisol also limits the migration of immune cells to sites of inflammation, thereby reducing the intensity of the inflammatory response. These combined actions help maintain the delicate balance between protecting against pathogens and preventing damage caused by excessive inflammation.

In addition, aldosterone, another adrenal hormone, is involved in regulating the body's electrolyte and water balance, which may have indirect implications for immunity by modulating the cellular environment in which immune responses occur. Similarly, DHEA, although its specific role in the immune response is less understood, is known for its effects on modulating inflammation and regulating the cellular immune response.

Thus, cortisol and other adrenal hormones do more than simply manage stress; they play a crucial role in maintaining immune homeostasis, ensuring effective defense against aggression while avoiding excessive responses that could cause tissue damage. Understanding these immunomodulatory mechanisms in detail offers promising prospects for the development of new therapeutic strategies aimed at regulating the immune response in a wide range of diseases, from chronic inflammation to autoimmune disorders.

Analysis of age-related immune changes and their influence on overall health

With aging, the immune system undergoes a series of profound and complex transformations, a process often referred to as immunosenescence. This evolution encompasses a panoply of changes that affect the immune response as a whole, ranging from decreased efficiency of immune cells to altered inflammatory regulatory mechanisms. These changes have important implications for the overall health of older individuals, influencing their resistance to infection, their ability to fight disease and their response to medical therapies.

One of the major characteristics of immunosenescence is a progressive decline in immune function, including an alteration in the ability of immune cells to recognize and eliminate pathogens. Immune system cells, such as T and B lymphocytes, become less effective with age, making individuals more vulnerable to viral and bacterial infections.

In addition, immunosenescence is associated with changes in the body's inflammatory response. While an inflammatory response is normally beneficial for fighting infection and healing damaged tissue, it often becomes dysfunctional in the elderly. This leads to increased levels of pro-inflammatory cytokines, a phenomenon known as low-grade inflammation, which can contribute to the development of chronic diseases such as cardiovascular disease, type 2 diabetes and neurodegenerative diseases.

At the same time, immunosenescence is also associated with an increased prevalence of autoimmune diseases in the elderly. This increased susceptibility results in part from disturbances in immune tolerance, leading to inappropriate autoimmune responses against body tissues.

In sum, analysis of age-related immune changes highlights the complexity of immunosenescence and its impact on the overall health of older individuals. Understanding these changes is essential for developing medical interventions and public health strategies aimed at mitigating the adverse effects of immunosenescence and improving the quality of life of the elderly.

Impact of exogenous corticoids on the immune response and their role in the treatment of age-related inflammatory diseases

Exogenous corticosteroids, such as prednisone and dexamethasone, represent fundamental tools in the therapeutic arsenal for treating inflammatory diseases in the elderly, such as rheumatoid arthritis and chronic obstructive pulmonary disease. Their mechanism of action is based on their ability to mimic the action of endogenous cortisol, the key hormone produced by the adrenal glands, in suppressing inappropriate inflammatory responses.

These exogenous corticoids act by inhibiting the production of pro-inflammatory cytokines and blocking the activity of immune cells involved in inflammatory processes. As a result, they are effective in alleviating symptoms and slowing the progression of inflammatory diseases, thus improving the quality of life of affected elderly people.

However, prolonged use of exogenous corticosteroids presents significant clinical challenges. While they may relieve inflammatory symptoms, they can also compromise the body's immune function, increasing the risk of opportunistic infections. This immune suppression can be of particular concern in the elderly, who already have weakened immune function due to immunosenescence.

In addition, exogenous corticosteroids can have undesirable side effects, such as osteoporosis, high blood pressure and weight gain, which can aggravate pre-existing health problems in the elderly.

In summary, although exogenous corticosteroids play a crucial role in the treatment of age-related inflammatory diseases, their use must be carefully balanced to maximize therapeutic benefits while minimizing potential health risks. A thorough understanding of their impact on the immune response and their clinical implications is essential for the effective management of inflammatory

diseases in the elderly.

Chapter 4: Regulation of blood pressure by adrenal hormones

This chapter takes a deep dive into the sophisticated mechanisms underlying the interaction between adrenal hormones, such as cortisol and aldosterone, and blood pressure regulation. We'll explore how these hormones interact with other key systems in the body, such as the renin-angiotensin-aldosterone system and the autonomic nervous system, to orchestrate a dynamic balance of blood pressure.

Take cortisol, for example, known for its varied effects on metabolism and the immune system. Studies have shown that cortisol can also directly influence blood pressure by altering the sensitivity of blood vessels to other vasoconstrictor and vasodilator hormones. This complex interaction can have a significant impact on blood pressure, especially when altered by factors such as aging or adrenal disorders.

By taking a closer look at the changes associated with aging and adrenal dysfunction, we can better understand how these alterations can influence blood pressure. For example, a decrease in aldosterone production due to aging can lead to reduced sodium retention and blood volume, thus contributing to a drop in blood pressure. Conversely, excessive cortisol secretion, as seen in Cushing's syndrome, can lead to hypertension due to its effects on blood vessel sensitivity and sodium retention.

In addition, we will explore the implications of these changes for the development of cardiovascular disease in the elderly. Research has shown that elevated cortisol levels can promote atherosclerosis, while disturbances in aldosterone regulation can contribute to high blood pressure, a major risk factor for cardiovascular disease.

By better understanding these complex mechanisms, we will be better equipped to develop more precise and personalized therapeutic interventions aimed at preventing and treating blood pressure problems in the elderly. This could include approaches such as modulation of adrenal hormones by drugs or dietary interventions, offering new prospects for improving cardiovascular health in aging individuals.

Chapter 5: Therapeutic approaches and future prospects

This chapter offers a deep dive into therapeutic options for treating disorders associated with impaired adrenal function, highlighting the use of corticosteroids and emerging approaches. We'll also explore recent advances in bioinorganic research to better understand underlying mechanisms and their relevance to the development of innovative treatments. Finally, we will examine the promising prospects of personalized medicine and targeted therapies to improve the health of the elderly.

In the field of bioinorganics, research has intensified to elucidate the complex interactions between adrenal hormones, such as cortisol, and essential enzymes like superoxide dismutase (SOD) and catalase. These enzymes are key players in the neutralization of reactive oxygen species (ROS), potentially toxic molecules produced during cellular metabolism. Cortisol, through its role as a hormonal regulator, directly influences the expression and activity of these enzymes, thus modulating the level of oxidative stress in the body.

For example, studies have shown that high levels of cortisol can stimulate the synthesis of SOD and catalase, boosting cellular defences against ROS. However, adrenal dysfunction, characterized by impaired cortisol production, can upset this balance. Decreased cortisol production can lead to reduced SOD and catalase activity, increasing susceptibility to oxidative damage and contributing to the development of age-related diseases such as cardiovascular and neurodegenerative diseases.

These findings pave the way for innovative therapeutic approaches aimed at restoring the disturbed redox balance in the elderly. For example, therapies aimed at selectively regulating SOD and catalase activity could offer new perspectives in the management of diseases associated with oxidative stress and adrenal dysfunction.

In summary, bioinorganics research is broadening our understanding of the underlying mechanisms of adrenal disorders and offering promising avenues for the development of personalized, targeted treatments. Combined with other therapeutic advances and personalized medicine, this approach offers exciting prospects for improving the health and quality of life of the elderly.

Conclusion

The conclusion of this research work underlines the vital importance of adrenal hormones in human health, particularly in the elderly. Through the various chapters explored, we have highlighted the complex and interconnected roles of cortisol, aldosterone and other adrenal hormones in the regulation of stress, metabolism, immune response and blood pressure.

Firstly, we examined in detail the mechanisms by which these hormones influence these vital physiological processes, notably by regulating inflammatory responses, sugar and fat metabolism, and the stress response. We also explored the consequences of adrenal dysfunction, highlighting its link to various age-related diseases, such as cardiovascular disease, diabetes and cognitive impairment.

Secondly, by exploring current and emerging therapeutic approaches, we have highlighted the challenges and opportunities in treating disorders associated with impaired adrenal function. The use of exogenous corticosteroids and advances in bioinorganic research offer promising prospects for improving the quality of life of the elderly while minimizing undesirable side effects.

Finally, by looking to the future of personalized medicine and targeted therapies, we are paving the way for a more precise and individualized approach to the management of hormonal disorders associated with aging. By better understanding the underlying mechanisms and developing innovative treatments, we can hope to promote the health and well-being of the elderly more effectively and sustainably.

In conclusion, this work highlights the crucial importance of adrenal hormones in maintaining human health, particularly in the elderly. By better understanding their role and complex interactions, we can hope to develop more effective therapeutic approaches and more targeted prevention strategies to improve the quality of life of the elderly and promote healthy aging for all.

Glossary :

Cortisol: Steroid hormone produced by the adrenal cortex, often referred to as the "stress hormone" because of its role in the stress response and regulation of carbohydrate metabolism.

Aldosterone: Steroid hormone produced by the adrenal cortex, involved in regulating electrolyte balance and blood pressure by promoting sodium reabsorption and potassium excretion in the kidneys.

Adrenal glands: Pairs of endocrine organs located at the top of each kidney, made up of the medullary adrenal gland (responsible for producing adrenaline and noradrenaline) and the adrenal cortex (responsible for producing corticosteroid hormones).

Glucocorticoids: A class of steroid hormones produced by the adrenal cortex, including cortisol, with anti-inflammatory, immunosuppressive and metabolic effects.

Mineralocorticoids: A class of steroid hormones produced by the adrenal cortex, including aldosterone, which play a role in regulating electrolyte balance and blood pressure.

Exogenous corticoids: Synthetic steroid hormones that mimic the action of cortisol naturally produced by the adrenal glands. They are often used as anti-inflammatory drugs.

Immunosenescence: The aging process of the immune system, characterized by a decline in its function and effectiveness, making individuals more vulnerable to infections and autoimmune diseases.

Cortisol: Steroid hormone produced by the adrenal glands in response to stress. It regulates various physiological processes, including immune response and metabolism management.

Inflammatory response: Immune system reaction to aggression, characterized by increased blood flow, accumulation of immune cells and release of chemical mediators to fight pathogens or repair damaged tissue.

Immunomodulation: The process of regulating the immune response, adjusting the activity of the immune system to maintain homeostasis and respond adaptively to external stimuli.

Adrenal: Also known as the adrenal gland, this is a small gland located above each kidney that produces hormones essential for regulating stress, metabolism and other bodily functions.

Cortisol: Also known as the stress hormone, cortisol is a hormone produced by the adrenal glands in response to stress. It plays a crucial role in regulating metabolism, immune response and stress management.

Orthostatic hypotension: a sudden drop in blood pressure when moving from a sitting or lying to a standing position, which can lead to dizziness, lightheadedness or even fainting.

Metabolism: All the biochemical processes that take place in the body to sustain life, including digestion, respiration, body temperature regulation and energy production.

Immune function: The ability of the immune system to protect the body against pathogens such as bacteria, viruses and cancer cells.

Cardiovascular diseases: diseases of the heart and blood vessels, such as hypertension, coronary heart disease and stroke, which are among the leading causes of death worldwide.

Type 2 diabetes: A form of diabetes characterized by insulin resistance and insufficient insulin production by the pancreas, often associated with obesity and a sedentary lifestyle.

Cognitive disorders: Alterations in cognitive function, such as memory, attention, language and executive functions, which may be associated with normal aging or with diseases such as Alzheimer's.

Adrenal hormones: These hormones are produced by the adrenal glands, located above the kidneys, and include cortisol, aldosterone, and catecholamines such as adrenalin and noradrenalin.

Cortisol: Also known as the stress hormone, cortisol is a steroid hormone produced by the adrenal glands in response to stress. It plays an important role in regulating metabolism, the immune system and the stress response.

Aldosterone: This hormone is produced by the adrenal glands and plays an essential role in regulating blood pressure and volume by controlling the balance of sodium and potassium in the body.

Blood pressure: The force exerted by blood against the artery walls. It is measured in millimeters of mercury (mmHg) and is made up of two figures: systolic pressure (the maximum pressure when the heart contracts) and diastolic pressure (the minimum pressure when the heart relaxes).

Renin-Angiotensin-Aldosterone System (RAAS): A complex hormonal system that regulates blood pressure and volume by controlling the secretion of the enzyme renin, which converts angiotensinogen to angiotensin I, and by regulating the release of aldosterone.

SOD (Superoxide Dismutase): An enzyme involved in the neutralization of free radicals, in particular anionic superoxide, which is a highly reactive free radical.

Catalase: An enzyme that breaks down hydrogen peroxide into water and oxygen, playing a crucial role in cellular defense against oxidative damage.

Oxidative stress: An imbalance between the production of free radicals and the ability of the body's antioxidant defense systems to neutralize them, leading to cell damage.

Adrenal dysfunction: Impaired function of the adrenal glands, leading to inadequate production of adrenal hormones such as cortisol and aldosterone.

References :

Buttgereit, F., Burmester, G. R., & Brand, M. D. (2020). Bioenergetics of immune functions: fundamental and therapeutic aspects. Immunological Reviews, 295(1), 174-186.

Franceschi, C., Garagnani, P., Parini, P., Giuliani, C., & Santoro, A. (2018). Inflammaging: a new immune-metabolic viewpoint for age-related diseases. Nature Reviews Endocrinology, 14(10), 576-590.

Prattichizzo, F., De Nigris, V., Spiga, R., Mancuso, E., La Sala, L., Antonicelli, R., ... & Ceriello, A. (2018). Inflammageing and metaflammation: The yin and yang of type 2 diabetes. Ageing Research Reviews, 41, 1-17.

Sapey, E., Greenwood, H., Walton, G., Mann, E., Love, A., Aaronson, N., ... & Lord, J. M. (2014). Phosphoinositide 3-kinase inhibition restores neutrophil accuracy in the elderly: toward targeted treatments for immunosenescence. Blood, 123(2), 239-248.

Torre, L. A., Bray, F., Siegel, R. L., Ferlay, J., Lortet-Tieulent, J., & Jemal, A. (2015). Global cancer statistics, 2012. CA: A Cancer Journal for Clinicians, 65(2), 87-108.

McEwen, B. S. (2002). The neurobiology of stress: from serendipity to clinical relevance. Brain Research, 886(1-2), 172-189.

Ouanes, S., & Popp, J. (2019). High Cortisol and the Risk of Dementia and Alzheimer's Disease: A Review of the Literature. Frontiers in Aging Neuroscience, 11, 43.

Shibao, C., Lipsitz, L. A., Biaggioni, I., & American Society of Hypertension Writing Group (2013). Evaluation and treatment of orthostatic hypotension. Journal of the American Society of Hypertension, 7(4), 317-324.

Soysal, P., Veronese, N., Arik, F., Kalan, U., Smith, L., Isik, A. T., & Stubbs, B. (2019). Orthostatic hypotension and health outcomes: An umbrella review of observational studies. European Journal of Internal Medicine, 61, 26-34.

Xu, W. L., Atti, A. R., Gatz, M., Pedersen, N. L., & Johansson, B. (2011). Midlife overweight and obesity increase late-life dementia risk: a population-based twin study. Neurology, 76(18), 1568-1574.

Barrett, K. E., Barman, S. M., Boitano, S., & Brooks, H. L. (2012). Ganong's Review of Medical Physiology. McGraw-Hill Medical.

Chrousos, G. P. (1995). The Hypothalamic-Pituitary-Adrenal Axis and Immune-Mediated Inflammation. New England Journal of Medicine, 332(20), 1351-1362.

Kenney, W. L., Wilmore, J. H., & Costill, D. L. (2012). Physiology of Sport and Exercise. Human Kinetics.

Miller, W. L. (2017). Disorders in the Production of Adrenal Hormones. In J. L. Jameson, & L. J. De Groot (Eds.), Endocrinology: Adult and Pediatric (7th Edition) (pp. 1654-1674). Elsevier.

Weitzman, R. E., & Ference, B. A. (2020). Prednisone and Prednisolone. In StatPearls. Treasure Island (FL): StatPearls Publishing.

Fernández-Real, J. M., & Ricart, W. (2003). Insulin resistance and chronic cardiovascular inflammatory syndrome. Endocrine Reviews, 24(3), 278-301.

Michaud, M., Balardy, L., Moulis, G., Gaudin, C., Peyrot, C., Vellas, B., ... & Nourhashemi, F. (2013). Proinflammatory cytokines, aging, and age-related diseases. Journal of the American Medical Directors Association, 14(12), 877-882.

Papaconstantinou, J. (2009). The role of signaling pathways of inflammation and oxidative stress in development of senescence and aging phenotypes in cardiovascular disease. Cells, 8(12), 1383.

Brownlee, M. (2001). Biochemistry and molecular cell biology of diabetic complications. Nature, 414(6865), 813-820.

Esterbauer, H., Schaur, R. J., & Zollner, H. (1991). Chemistry and biochemistry of 4-hydroxynonenal, malonaldehyde and related aldehydes. Free Radical Biology and Medicine, 11(1), 81-128.

Butterfield, D. A., & Lauderback, C. M. (2002). Lipid peroxidation and protein oxidation in Alzheimer's disease brain: potential causes and consequences involving amyloid β-peptide-associated free radical oxidative stress. Free Radical Biology and Medicine, 32(11), 1050-1060.

Guyton, A.C., & Hall, J.E. (2015). Textbook of Medical Physiology. Philadelphia, PA: Saunders.

Becker, K.L. (2001). Principles and Practice of Endocrinology and Metabolism. Philadelphia, PA: Lippincott Williams & Wilkins.

Ross, A.C., Caballero, B., Cousins, R.J., et al. (2014). Modern Nutrition in Health and Disease. Baltimore, MD: Lippincott Williams & Wilkins.

Rainey, W.E., & Nakamura, Y. (2008). Regulation of the Adrenal Cortex. Endocrine Development, 20, 99-116.

Bornstein, S.R., Allolio, B., Arlt, W., et al. (2016). Diagnosis and Treatment of Primary Adrenal Insufficiency: An Endocrine Society Clinical Practice Guideline. The Journal of Clinical Endocrinology & Metabolism, 101(2), 364-389.

yes **I want** morebooks!

Buy your books fast and straightforward online - at one of world's fastest growing online book stores! Environmentally sound due to Print-on-Demand technologies.

Buy your books online at
www.morebooks.shop

Kaufen Sie Ihre Bücher schnell und unkompliziert online – auf einer der am schnellsten wachsenden Buchhandelsplattformen weltweit! Dank Print-On-Demand umwelt- und ressourcenschonend produziert.

Bücher schneller online kaufen
www.morebooks.shop

info@omniscriptum.com
www.omniscriptum.com

Printed by Books on Demand GmbH, Norderstedt / Germany